WORKBOOK
Working for over 25 YEARS WITH Cambridge Assessment International Education
Cambridge Assessment International Education
Endorsed for learner support
Cambridge IGCSE™ and O Level
Geography
Second edition
Paul Guinness
Garrett Nagle
T0273463
HODDER EDUCATION

Photo credits

The Publishers would like to thank the following for permission to reproduce copyright material.

Paul Guinness: p. 52, p. 57

Garrett Nagle: p. 26, A and B, p.34, p.40, p.91, p.92, p.93, all, p.94

Acknowledgements

Every effort has been made to trace all copyright holders, but if any have been inadvertently overlooked, the Publishers will be pleased to make the necessary arrangements at the first opportunity.

Although every effort has been made to ensure that website addresses are correct at time of going to press, Hodder Education cannot be held responsible for the content of any website mentioned in this book. It is sometimes possible to find a relocated web page by typing in the address of the home page for a website in the URL window of your browser.

Hachette UK's policy is to use papers that are natural, renewable and recyclable products and made from wood grown in well-managed forests and other controlled sources. The logging and manufacturing processes are expected to conform to the environmental regulations of the country of origin.

The questions, example answers, marks awarded and/or comments that appear in this book were written by the authors. In examination, the way marks would be awarded to answers like these may be different.

Orders: please contact Hachette UK Distribution, Hely Hutchinson Centre, Milton Road, Didcot, Oxfordshire, OX11 7HH. Telephone: +44 (0)1235 827827. Email education@hachette.co.uk
Lines are open from 9 a.m. to 5 p.m., Monday to Friday. You can also order through our website: www.hoddereducation.com

First published in 2015 by
Hodder Education,
An Hachette UK Company
Carmelite House, 50 Victoria Embankment, London EC4Y 0DZ
This second edition published 2018

Impression number 10 9 8 7 6 5

Year 2021

Cover photo Shutterstock/Joachim Wendler

Typeset by Aptara, Inc.

Printed and bound in the UK

A catalogue record for this title is available from the British Library.

ISBN: 978 1 5104 2138 7

Contents

Introduction

Welcome to the Cambridge IGCSE and O Level Geography Workbook. The aim of this Workbook is to provide you with further opportunity to practise the skills you have acquired through using the IGCSE and O Level Geography textbook. It is designed to complement the third edition of the textbook and to provide additional exercises to help you in preparation for your examinations. This Workbook covers the content of the Cambridge IGCSE/O Level syllabuses.

The chapters in this Workbook reflect the topics in the textbook.

There is no set way to approach using this Workbook. You may wish to use it to supplement your understanding of the different topics as you work through each chapter of the textbook, or you may prefer to use it to reinforce your skills in dealing with particular topics as you prepare for examination. The Workbook is intended to be sufficiently flexible to suit whatever you feel is the best approach for your needs.

1 Population and settlement

1.1 Population dynamics

1 Look at Figure 1.3 on page 3 of the textbook. When did the world's population reach:

a 1 billion? .... **b** 5 billion? .... **c** 7 billion? ....

2 When is the world's population forecast expected to reach 8 billion? ....

3 In your textbook, look at Table 1.1 on page 3, and read the paragraph on page 3 commenting on Table 1.1.

a How many people were born in 2016? ....

b How many people died in 2016? ....

c By how much did the world's population increase in 2016? ....

4 When was the highest rate of global population growth? ....

5 Complete the table to match the following world regions to their proportion of the world's population: Africa, Asia, Europe, Latin America/Caribbean, North America, Oceania.

Region	% of world population, 2016 (rounded)
	59.3
	16.2
	10.0
	8.5
	4.8
	1.5

6 List the five largest countries in terms of population in 2016.

1 ....

2 ....

3 ....

4 ....

5 ....

7 On the diagram below, insert the five labels required to complete the diagram.

Natural change

8 Define:

a 'birth rate'

....

b 'death rate'

....

9 Which world regions have the highest and lowest birth rates?

Highest ....

Lowest ....

10 If a country has a birth rate of 32/1000 and a death rate of 8/1000, what is its rate of natural increase?

....

11 Name the five stages of the demographic transition model.

Stage 1 ....

Stage 2 ....

Stage 3 ....

Stage 4 ....

Stage 5 ....

12 Where and why in the demographic transition model is the rate of population growth highest?

....

....

....

....

13 Define 'total fertility rate'.

....

....

14 Give one reason why the total fertility rate is a better/more detailed measure of fertility than the birth rate.

....

....

15 How can the infant mortality rate influence the level of fertility in a country?

....

....

....

16 a What type of graph is Figure 1.10 on page 7 of your textbook? ....

b What name is given to the slanted line on the graph? ....

17 Describe the correlation between the total fertility rate and the percentage of girls enrolled in secondary education, illustrated by Figure 1.10.

....

....

....

18 Define 'life expectancy at birth'.

....

....

19 Which world regions have the highest and lowest life expectancy figures?

Highest ....

Lowest ....

20 In which type of country do infectious diseases such as malaria and tuberculosis kill many people?

....

21 What is the 'demographic divide'?

....

....

....

22 Where in the world is HIV/AIDS most prevalent? ....

23 Describe two factors responsible for high rates of HIV/AIDS.

1 ....

....

2 ....

....

24 Define 'optimum population'.

....

....

....

25 Give two reasons why the world as a whole might be considered to be overpopulated.

1 ....

....

2 ....

....

26 Define 'population policy'.

....

....

27 What is the name given to a population policy that:

a promotes large families? ....

b aims to reduce population growth? ....

28 How effective was China's one-child policy in reducing fertility?

....

....

....

....

....

....

29 What problems did the one-child policy create in China?

....

....

....

....

....

....

30 Why has France, along with a number of other developed countries, taken measures to encourage fertility?

....

....

....

....

1.2 Migration

Look at Figure 1.31 on page 20 of the textbook.

1 Define 'migration'.

2 Add labels to the diagrams below to show four push factors and four pull factors.

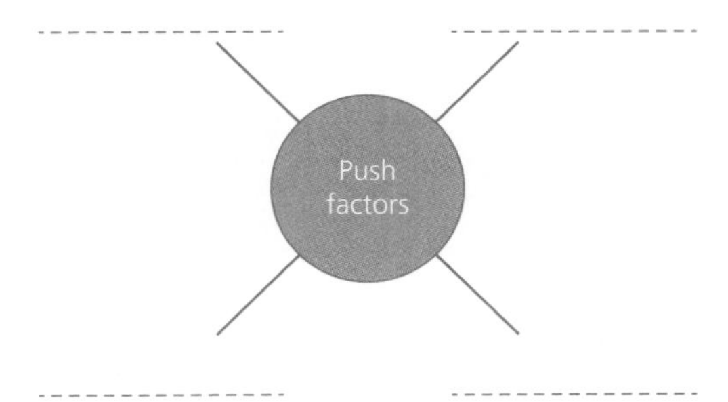

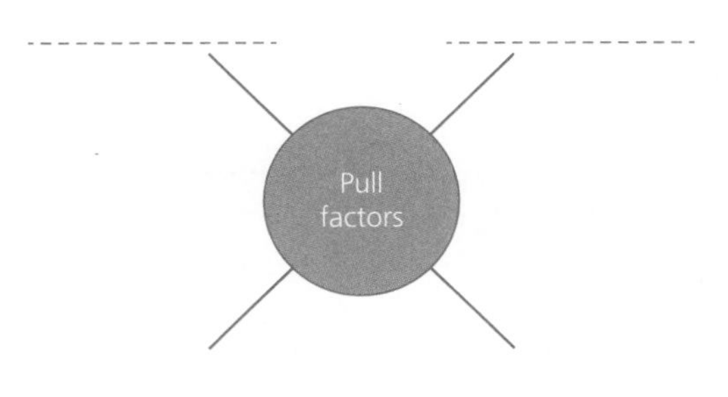

3 Give an example to explain voluntary migration.

4 Give an example to explain involuntary (forced) migration.

5 **a** How many forcibly displaced people are there worldwide? ....

b How many of these people are classed as refugees? ....

c Define 'refugee'.

6 Suggest why more international migrants move from the developing world to the developed world rather than in the reverse direction.

....

....

....

....

7 Why is population movement within countries at a much higher level than movements between countries?

....

....

....

8 In which country is the largest-ever internal migration currently taking place?

....

9 Define 'rural depopulation'.

....

....

10 Give two causes of rural depopulation.

1 ....

....

2 ....

....

11 What may eventually happen to a rural settlement after a long period of depopulation?

....

....

12 Define 'counterurbanisation'.

....

....

13 Why has counterurbanisation occurred in so many developed countries?

....

....

....

14 Fill in two examples of the impact of international migration in each of the six cells in the table below.

Impact on countries of origin	Impact on countries of destination	Impact on migrants themselves
Positive 1 .... 2	Positive 1 .... 2	Positive 1 .... 2 ....
Negative 1 .... 2	Negative 1 .... 2	Negative 1 .... 2 ....

15 Define 'remittances'.

....

....

16 How high were global remittance flows in 2016? ....

17 Give two examples of the benefits of remittances to developing countries.

1 ....

2 ....

18 Which is the highest source of funding for developing countries — remittances or official international aid?

....

1.3 Population structure

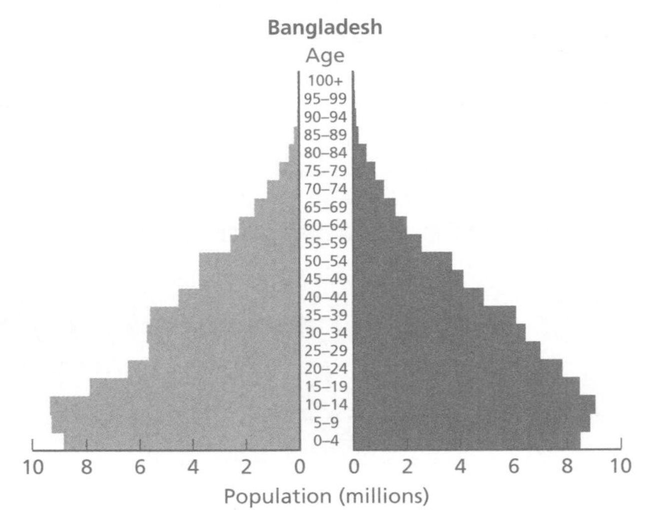

Bangladesh: the population pyramid for 2013

1 Define 'population structure'.

....

....

2 On the figure above:

a add the labels 'females' and 'males' to the appropriate sides of the population pyramid.

b draw horizontal lines and provide labels to divide the pyramid into the young dependent population, the economically active population and the elderly dependent population.

3 How many years of age are represented by each bar on a population pyramid? ....

4 Suggest why the 0–4 and 5–9 bars on the population pyramid for Bangladesh are not as wide as the 10–14 bar.

....

....

5 The data in the figure above are shown in millions [absolute data]. What is the alternative way of showing data on a population pyramid?

....

6 Bangladesh is in which stage of demographic transition? ....

7 What will happen to the shape of Bangladesh's population pyramid as it moves to the next stage of demographic transition?

....

....

....

8 In which stages of demographic transition are:

a Japan? ....

b Niger? ....

c the UK? ....

9 Why is the population structure of urban and rural areas in the same country sometimes markedly different?

....

....

....

....

....

10 Define the 'dependency ratio'.

....

....

11 What does a dependency ratio of 70 mean?

....

....

12 Is the dependency ratio generally higher in developed or developing countries?

....

13 Why is the dependency ratio an important factor for a country?

....

....

....

....

....

14 Give one limitation of the dependency ratio.

....

15 Give an example of a country with a high dependent population. ....

1.4 Population density and distribution

1 Explain the difference between population density and population distribution.

....

....

....

....

....

....

2 Explain the following terms:

a 'densely populated' ....

....

b 'sparsely populated' ....

....

3 Use your textbook to insert the population density figures for each world region in the table below.

Region	Population density (people per km^2) in 2016
World	
Africa	
Asia	
Latin America/Caribbean	
North America	
Europe	
Oceania	

4 How much higher is average population density in the less developed world compared with the more developed world?

....

5 Which world region has:

a the highest population density? ....

b the lowest population density? ....

6 Give three types of natural environment associated with very low population densities.

1 ....

2 ....

3 ....

7 Discuss the physical and economic factors that encourage high population density.

....

....

....

....

....

....

....

....

8 Give an example of one social factor and one political factor that could influence population distribution.

....

....

....

....

9 Complete the fact file below for the Canadian northlands — a sparsely populated region.

Location	
Population density	
Temperature	
Permafrost	
Economic activities	
Transport	
Settlement	

10 a Name the most densely populated region in the USA. ....

b List four major cities in this region.

....

....

c Give two reasons for this region's high population density.

....

....

1.5 Settlements and service provision

1 Define the terms:

a 'high-order services (or goods)'

....

....

....

b 'low-order services (or goods)'

....

....

....

2 Arrange these settlements in descending order of size: *city, conurbation, hamlet, town, village.*

....

3 Study the map of the Lozère district in France below. Refer to page 45 of the textbook for data on the settlements.

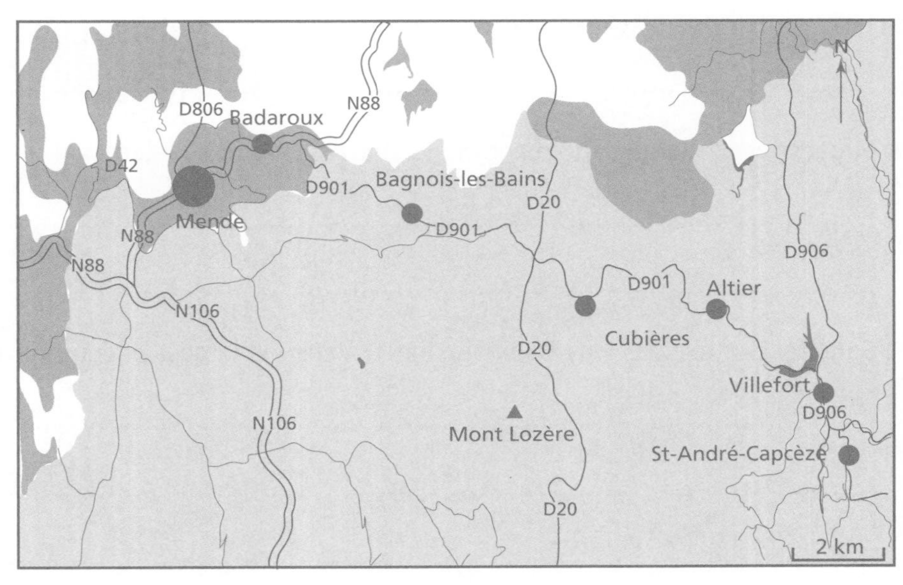

a i Describe the site of Mende.

....

....

ii Describe the site of St-André-Capcèze.

....

....

iii Describe the situation of Bagnois-les-Bains.

....

....

b i Suggest why Mende has the highest population in the region.

....

....

ii Suggest why St-André-Capcèze has such a small population.

....

....

iii Identify, and justify, a likely location for another large settlement in the area.

....

....

iv Suggest why a large settlement might not develop in this location.

....

....

v Suggest two reasons why Badaroux has the second largest population in the study area.

....

....

....

....

1.6 Urban settlements

Urbanisation trends

The distribution of the world's urban population by region (%)

Region	1950	2011	2050
Africa	4	11	21
Asia	33	52	53
Europe	38	15	9
Latin America	9	13	10
North America	15	8	6
Oceania	1	1	1

1 Using the data in the table, draw three sets of bar charts to show the changes in the distribution of the world's urban population by region, 1950, 2011 and 2050.

2 In which region is the most urban increase predicted for 2011–2050? ....

3 In which region is the share of urban growth predicted to increase most between 1950–2011 and 2011–2050? ....

4 In which regions is the share of urban growth falling?

....

5 Briefly outline three advantages of living in an urban area.

1 

2 ....

....

3 ....

....

6 Briefly outline three disadvantages of living in an urban area.

1 ....

....

2 ....

....

3 ....

....

Urban settlements in LICs, MICs and HICs

Percentage of total urban settlements

Population size	Low-income countries	Middle-income countries	High-income countries
Small settlements (<20,000)	73	55	22
Medium-sized settlements (20,000–1,000,000)	16	25	26
Large-sized settlements (over 1,000,000)	11	20	52

1 Compare the proportions of small settlements in LICs, MICs and HICs.

....

....

....

....

....

2 Compare the proportions of large settlements in LICs, MICs and HICs.

....

....

....

....

....

3 Explain two advantages of large settlements over small settlements.

1 ....

....

2 ....

....

4 Explain two advantages of small settlements over large settlements.

1 ....

....

2 ....

....

Urban growth in Shanghai

The growth of Shanghai, 1978–2015

Year	Population (million)
1978	11.04
1980	11.52
1985	12.33
1990	13.34
1995	14.14
2000	16.08
2005	18.90
2010	23.02
2015	24.15

1 On the grid provided, plot the data for Shanghai's population growth (1978 and 1980 have already been plotted for you).

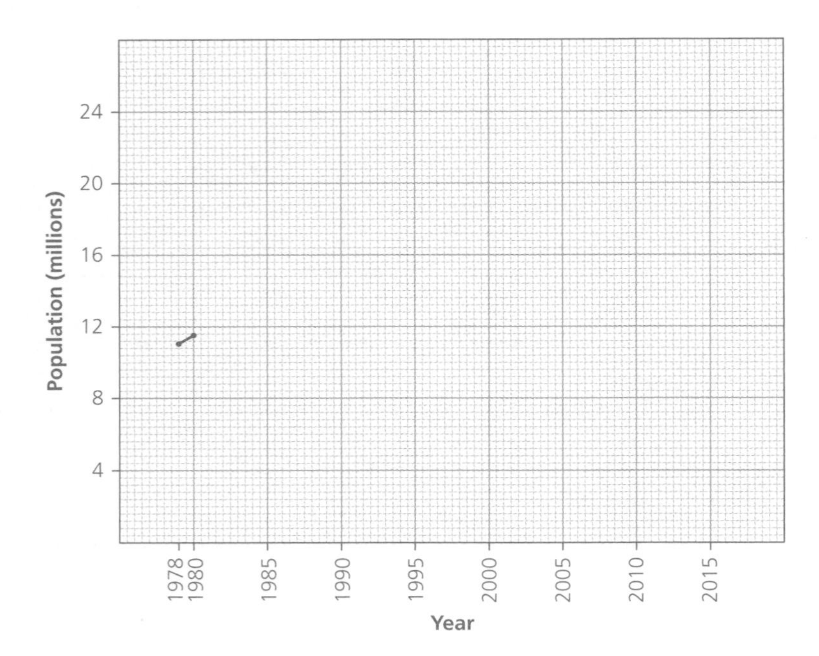

2 Describe Shanghai's urban growth between 1978 and 2015.

....

....

....

3 Suggest two contrasting reasons for urban growth around the world.

1 ....

2 ....

4 Suggest which one of your reasons (in question 3) is likely to explain urban growth in Shanghai. Justify your choice.

....

....

....

The diagram below shows a partially completed map of population density in Shanghai.

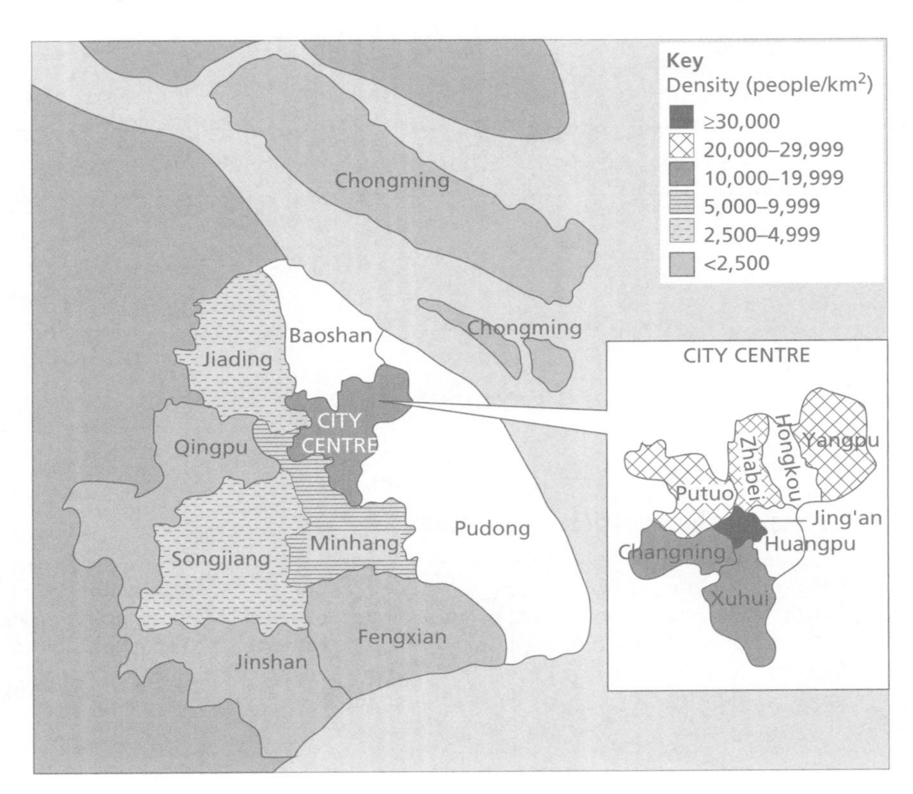

5 Complete the diagram by plotting the data for Pudong, Hongkou, Huangpu and Baoshan from the following table.

Administrative district	Population density (population/km²)
Pudong	4,523
Hongkou	34,501
Huangpu	32,190
Baoshan	7,465

The average population density in Shanghai is 3,809 people/km².

6 Describe the variations in population density in Shanghai.

....

....

....

....

7 Suggest two contrasting reasons for the population density in:

a Hongkou

1 ....

2 ....

b Chongming

1 ....

2 ....

Urban quality of life

Study the data in the table below and answer the questions that follow.

City	Current population in city in millions (not metro region)	Central area density (population/km²)	Projected growth 2010–2025 (pp hour)	% of the country's population living in the city	GDP pp (US$)	Life expectancy (years)	% under 20 years of age	Car ownership rates per 1,000	Daily water consumption (litres pp)	Annual CO_2 emissions (kg pp)
Hong Kong	7.0	22,193	7	–	45,080	82.5	20.1	59	371	5,800
New York	8.1	15,353	9	2.8	55,693	77.6	25.7	209	607	7,396
Shanghai	15.5	23,227	26	1.0	8,237	81.0	16.0	73	493	10,680
London	7.6	8,326	1	12.4	60,831	79.2	23.8	345	324	5,599
Mexico City	8.6	12,860	10	8.4	18,321	75.9	32.9	360	343	5,862
Johannesburg	3.9	2,208	3	8.1	9,259	51.0	34.6	201	378	5,025
Mumbai	11.7	45,021	44	0.9	1,871	68.1	36.3	36	90	371
São Paulo	10.4	10,326	11	5.8	12,021	70.8	31.0	368	185	1,123
Istanbul	12.7	20,128	12	17.8	9,368	72.4	32.1	139	155	2,720

Source: Based on *Urban Age*; pp = per pers

1 Identify the city with the largest population. ....

2 Which city has the highest population density? ....

3 Identify the city that is projected to have the largest population growth between 2010 and 2025.

....

4 Which city accounts for the largest percentage of its country's population? ....

5 a Identify the richest city in terms of GDP per person. ....

b Which is the poorest city in terms of GDP per person? ....

c Calculate the ratio of the richest city to the poorest city, in terms of GDP per person. ....

6 In which city is life expectancy:

a highest? ....

b lowest? ....

7 Which city has the highest proportion of young people (i.e. those aged under 20 years)?

....

8 Comment on levels of car ownership among the cities.

....

....

9 Complete the graph below and comment on the relationship between GDP in urban areas and CO_2 emissions.

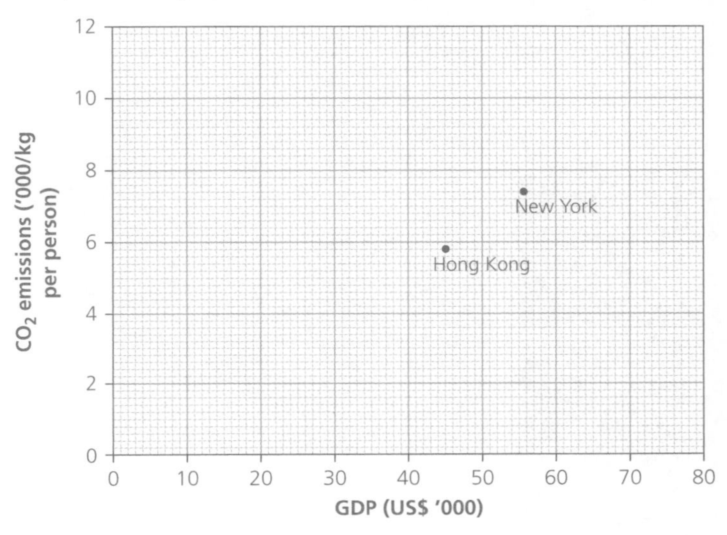

....

....

....

....

....

....

1.7 Urbanisation

The two photos below show a market town and a central business district.

Photo A

Photo B

1 Identify which photo shows:

a a market town ....

b part of a CBD ....

2 Describe the main characteristics of Photo A.

....

....

3 Describe the main characteristics of Photo B.

....

....

2 The natural environment

2.1 Earthquakes and volcanoes

Tectonics

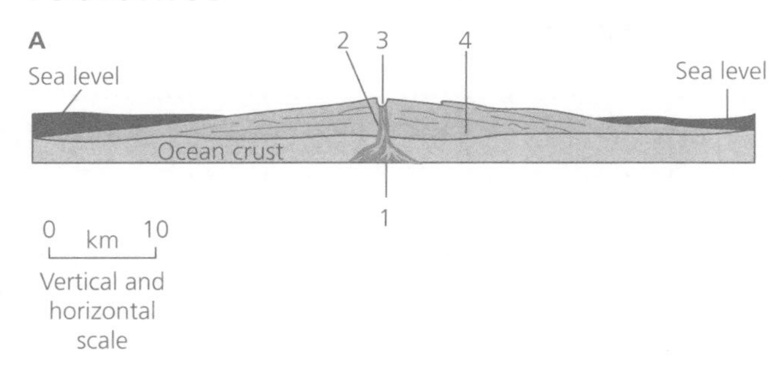

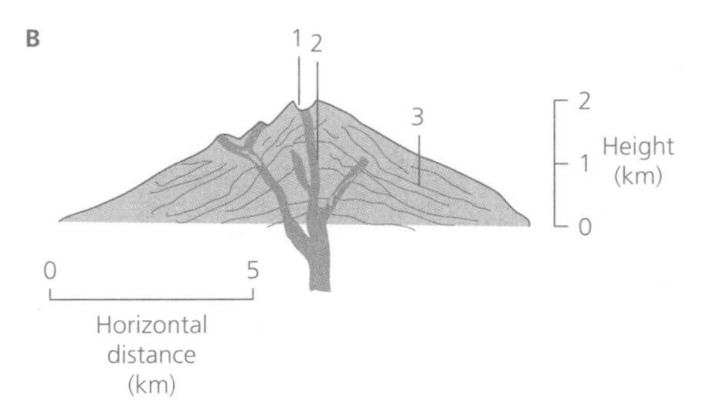

1 The figures show a cone volcano and a shield volcano.

a Identify both types of volcano.

A ....

B ....

b For each volcano, identify the numbered features.

A 1 ....

2 ....

3 ....

4 ....

B 1 ....

2 ....

3 ....

c Estimate the width and height of both types of volcano.

A ....

B ....

d Distinguish between magma and lava.

....

....

....

2 Describe the main characteristics of shield volcanoes and cone volcanoes.

....

....

....

The impact of volcanoes

Volcano	Year	Deaths	Major cause of deaths
Tambora, Indonesia	1815	92,000	Starvation
Krakatoa, Indonesia	1883	36,417	Tsunami
Mt Pelée, Martinique	1902	29,025	Ash flows
Nevado del Ruiz, Colombia	1985	25,000	Mudflows
Unzen, Japan	1792	14,300	Volcano collapse, tsunami
Laki, Iceland	1783	9,350	Starvation
Kelut, Indonesia	1919	5,110	Mudflows
Galunggung, Indonesia	1882	4,011	Mudflows
Vesuvius, Italy	1631	3,500	Mudflows, lava flows

Study the table above, which provides data on the most deadly volcanoes since 1600.

1 Which country has been most affected by deadly volcanoes? ....

2 Identify the most frequent hazard associated with volcanic eruptions. ....

3 Explain how volcanic eruptions may lead to starvation.

....

....

....

....

4 Distinguish between extinct, dormant and active volcanoes.

....

....

....

....

5 Suggest why the death toll from volcanoes is normally relatively low.

Earthquake impacts

The diagram below shows some of the characteristics of an earthquake and some impacts.

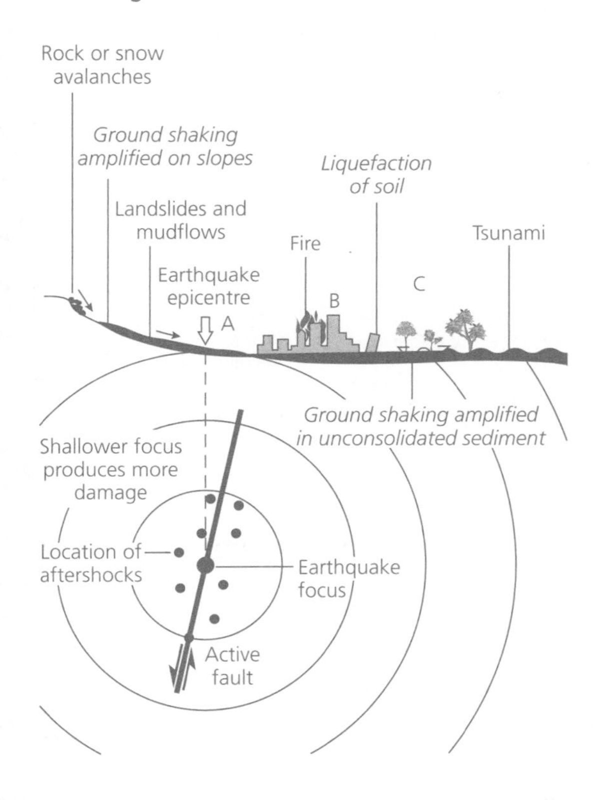

1 Define the terms 'epicentre' and 'focus' of an earthquake.

2 Explain why, for earthquakes of the same magnitude, shallow-focus earthquakes produce more damage than deep-focus earthquakes.

....

....

....

....

3 a Identify the location where damage from the earthquake is likely to be greatest — A, B or C.

....

b Justify your choice.

....

....

....

4 a Suggest what is meant by 'liquefaction of soil'.

....

b Why may liquefaction be a problem?

....

....

....

5 a What is an aftershock?

....

....

b In what ways can aftershocks be hazardous?

....

....

c Name one other hazard associated with earthquakes that is not shown in the diagram on page 29.

....

Living with volcanoes and earthquakes

1 Outline the ways of predicting volcanic eruptions.

....

....

....

....

2 Comment on the methods for predicting earthquakes.

....

....

....

....

....

3 Suggest ways in which volcanic eruptions can be managed.

....

....

....

....

....

4 Explain how earthquakes can be managed.

....

....

....

....

5 How does the management of volcanoes and earthquakes differ between MEDCs and LEDCs?

....

....

....

2.2 Rivers

Changes in sediment size

The diagram below shows changes in (a) sediment size and (b) sediment composition (make-up) in the Mississippi, USA.

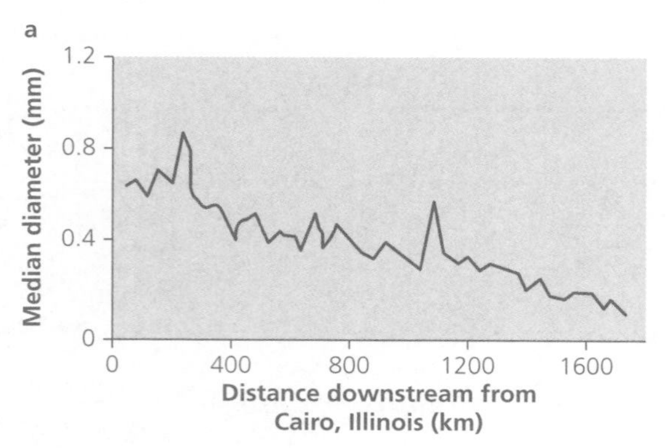

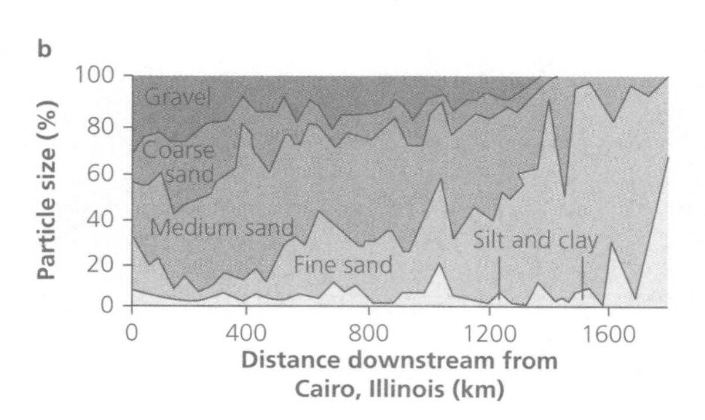

1 Describe the changes in median sediment diameter with distance downstream.

....

....

....

2 a Estimate the percentage for each sediment category at 800 km.

Gravel .... %

Coarse sand .... %

Medium sand .... %

Fine sand .... %

Silt and clay .... %

b Describe the changes in the percentage of gravel downstream.

....

....

c Describe the changes in the percentage of silt and clay downstream.

....

....

....

....

3 a How might the shape of particles be expected to change downstream?

....

....

b Suggest why sediment shape may change downstream.

....

....

4 a Name one other type of load carried by a river. ....

b Identify one rock type that may be a source of this load. ....

Waimakariri valley

Study the map of Arthur's Pass, New Zealand, on page 169 of the textbook.

1 a State the height of the highest point on the map. ....

b Identify the location and height of the lowest point on the map.

....

2 a Describe the characteristics of the BB Trail.

....

....

b Outline one hazard that may affect the tramping track.

....

....

3 a Describe three main characteristics of the Waimakariri River.

....

....

b Describe three characteristics of the stream in square 7936.

....

....

c Suggest reasons for the differences in the two rivers' characteristics.

....

....

....

....

....

Lowland rivers

The photograph below shows a river landscape in eastern Europe. The white areas away from the river channel are covered in snow. The sketch shows the same area.

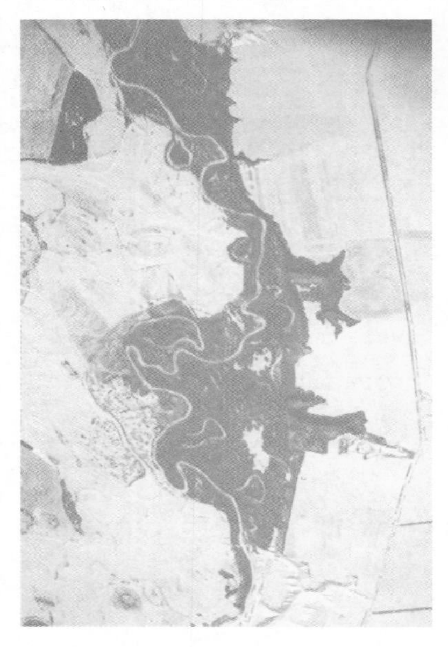

1 a Identify the landforms A and B:

A .... **B** ....

b Identify the main process happening at C (outer bank) and D (inner bank):

C .... **D** ....

c Suggest a likely feature to be found at the edge of the river channel at E. ....

2 a With the use of a diagram explain how oxbow lakes are formed.

b With the use of a diagram explain how feature E is formed.

Floods

1 Define 'flood'.

....

....

2 a Outline the disadvantages of floods.

....

....

b Comment on the advantages that floods bring.

....

....

3 a Describe the natural causes of floods.

....

....

b Outline the human factors that contribute to floods.

....

....

....

....

4 Explain why some floods are more intense than others.

....

....

....

....

....

....

2.3 Coasts

Mapwork

1 Study Figure 2.62 on page 132 of the textbook, which shows the Cape Peninsula in South Africa.

a Identify the location and height of the highest point on the map. ....

b Compare the relief of the slopes on the eastern side (Penguin Rocks) with those on the western side (Maclear Beach).

....

....

2 Study the map of Montego Bay, Jamaica, on page 46 of the textbook.

a Identify the vegetation feature in squares 5099 and 4800 (48100).

....

....

b Suggest reasons for the location of the Yacht Club at 503102.

....

....

....

3 Study the map of southwest Tenerife on page 276 of the textbook and the key below.

a Name three beaches for bathing.

....

....

....

b Where is it possible to go diving?

....

c Where is it possible to dock a boat?

....

....

Key

Swimming pool

Bathing beach

Diving

Haven, ship landing

Lighthouse

d Which beach is easier to reach: Playa de Masca or Playa de Barranco Seco? Explain your answer.

....

....

....

Depositional features

1 Draw a labelled diagram to show how and why longshore drift occurs.

2 a Identify the landforms A, B and C on the diagram below.

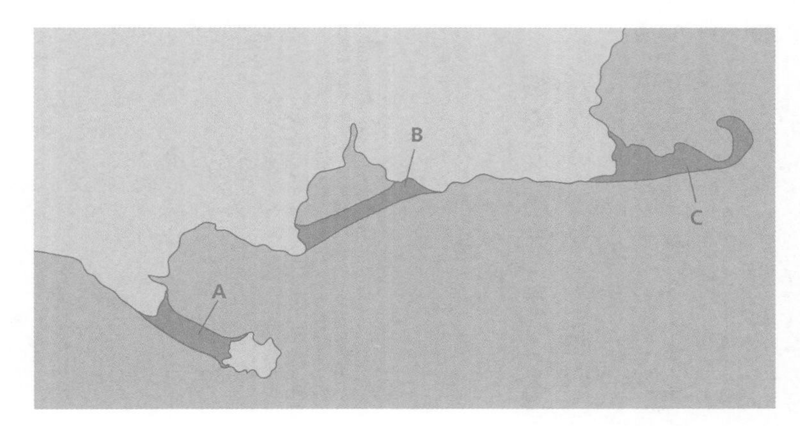

A ....

B ....

C ....

b Describe each of the three landforms.

A

....

....

B

....

....

C

....

....

3 Outline the conditions needed for the formation of landform C.

....

....

....

Coastal conflicts

The photograph below shows part of Barcelona port.

The tables below provide data for the number of cruise passengers arriving at Barcelona port, and the amount of load from containers.

Year	No. of passengers to Barcelona/year
1980s	77,000
1998	466,000
2002	843,000
2005	1,250,000
2010	2,400,000
2013	2,600,000

Year	Container ship load ('000 TEU – 20 foot equivalent)
1985	387
1995	683
2000	1,388
2005	2,100
2008	2,210
2010	1,931

(Source: Recent traffic dynamics in the European container port system, *Port Technology International*, Issue 58, 2013)

1 Complete the graph below, showing the number of passengers and the amount of load arriving in Barcelona.

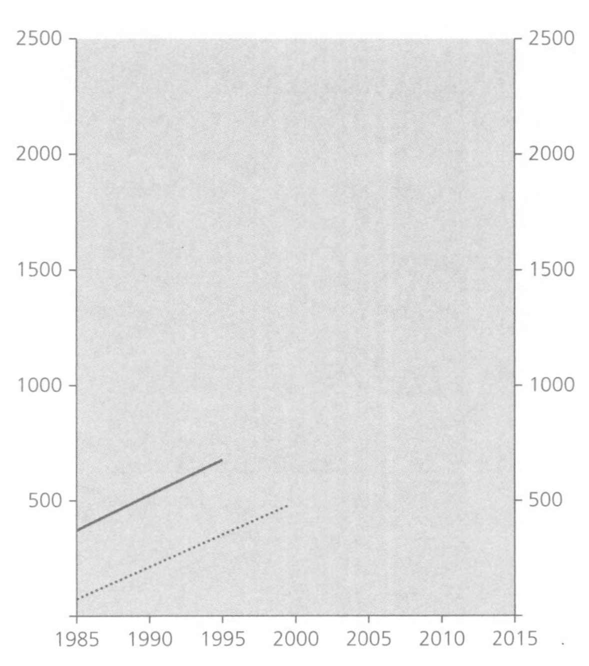

Graph to show changes in container trade and cruise ship passengers in Barcelona

2 Compare and contrast the changes in container load and number of cruise passengers.

....

....

....

....

3 Suggest some of the potential conflicts for users of the port of Barcelona, resulting from the increase in the number of cruise passengers and container load. The photograph on page 40 may suggest some potential conflicts.

....

....

....

....

....

....

2.4 Weather

Clouds

1 Study the diagram below, which shows different cloud types.

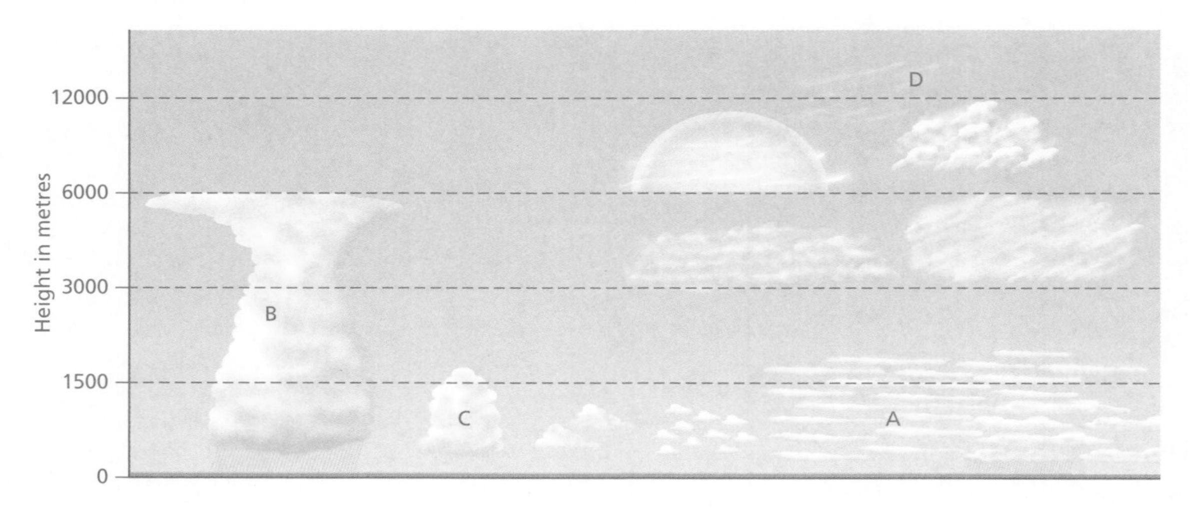

a Identify cloud types A, B, C and D.

A .... **C** ....

B .... **D** ....

b Describe the main characteristics of a stratus cloud and a cumulonimbus cloud.

....

....

....

....

c What are the highest clouds called? ....

d Which unit is used to measure cloud cover? ....

e Above which height are high clouds formed? ....

f At what height do low clouds generally have their base? ....

2 a What are high clouds formed from?

....

b What are low clouds mainly formed from?

....

c What does the word nimbus refer to in cloud development?

....

Wind rose

The diagram below shows a wind rose. The amount of time when there was no wind (i.e. calm conditions) is shown by the percentage in the centre of the diagram.

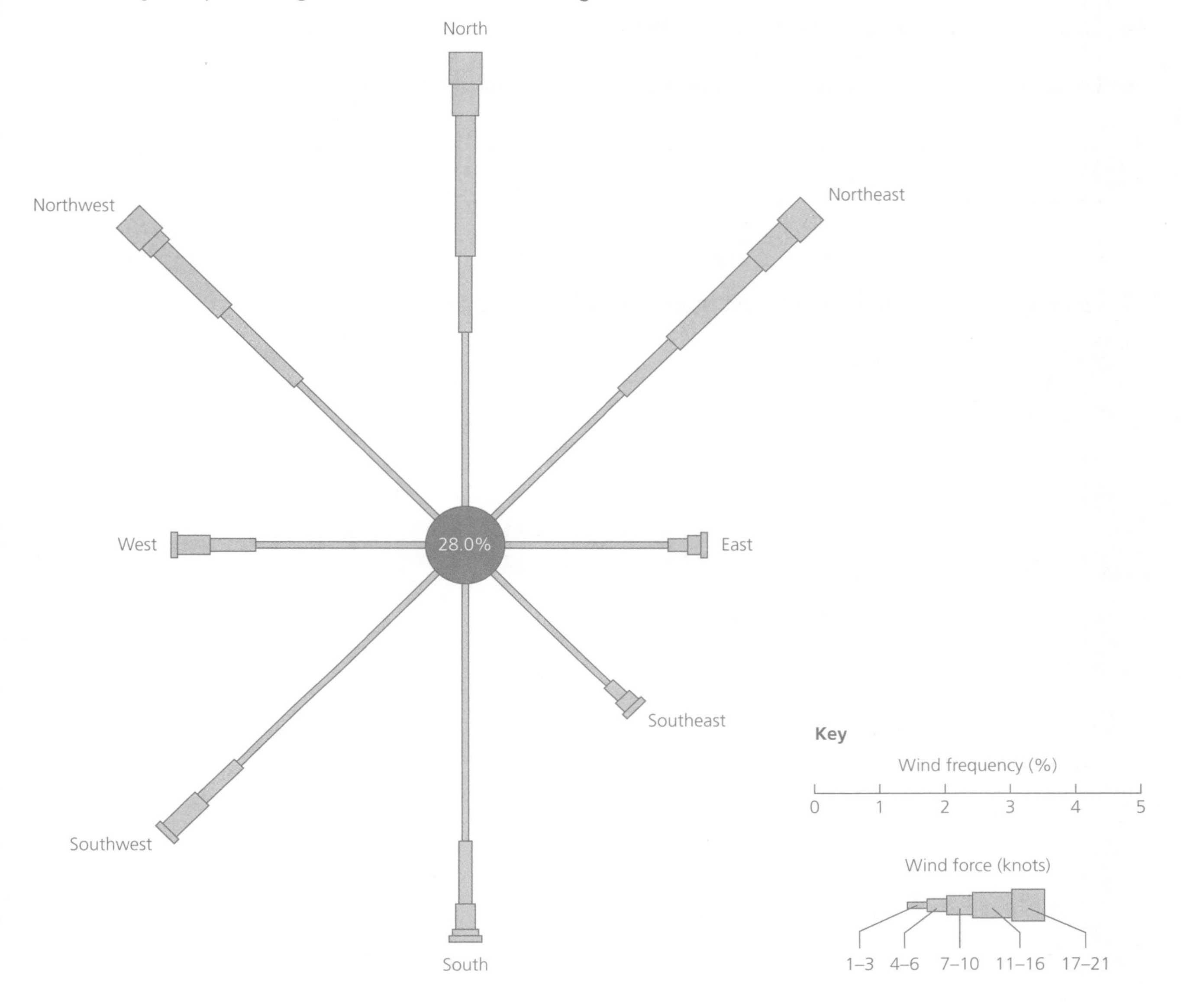

1 a Identify two features that a wind rose shows.

....

b State the direction from which the wind is most frequent on the wind rose diagram. ....

c State the term that describes the most frequent wind that an area experiences.

....

d State the proportion of time when conditions were calm. ....

e Compare the frequency and strength of winds from the north with those of winds from the south.

....

....

....

f From which direction is the highest proportion of winds of over 11 knots? ....

2.5 Climate and natural vegetation

Microclimate changes around a school

The diagram below shows part of a school grounds that was used for an investigation into 'variations in the school's microclimate'. The school is in the northern hemisphere.

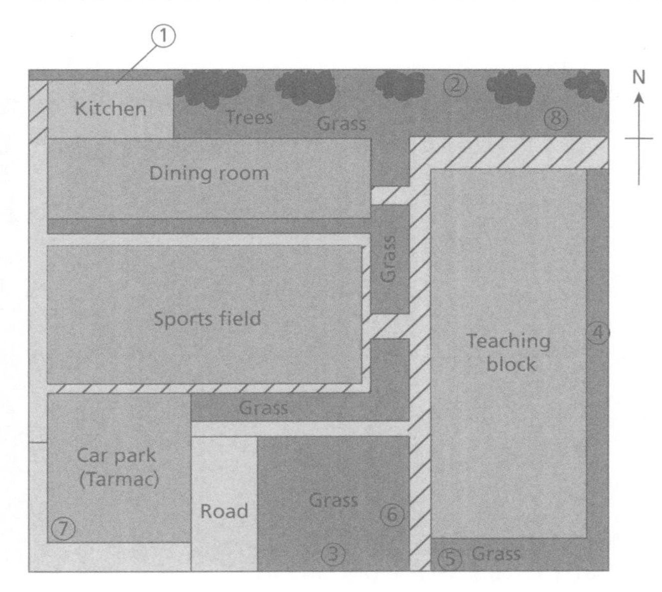

Suggest why:

1 Site 1 had the highest minimum temperature.

....

....

2 Site 6 was warmer than Site 4.

....

....

3 Site 5 was warmer than Site 8.

....

....

4 Site 2 had the lowest recorded rainfall.

....

....

5 Site 7 was warmer than Site 3.

....

....

6 Sites 3 and 7 had the highest rainfall.

....

....

....

Tropical rainforests

Look at Figure 2.98 on page 159 of the textbook, which shows the distribution of areas of tropical rainforest.

1 Describe the distribution of tropical rainforests.

....

....

....

2 Study Figure 2.99 on page 160 of the textbook, which shows characteristics of the vegetation found in tropical rainforests.

   a What is the maximum height of the vegetation? ....

   b Describe the characteristics of the vegetation found at the top of the rainforest.

   ....

   ....

   c Explain how this vegetation has adapted to conditions in the rainforest.

   ....

   ....

   ....

   ....

   d Comment on the vegetation found at ground level in tropical rainforests.

   ....

   ....

   ....

Hot deserts

Study Figure 2.105 on page 164 of the textbook, which shows the distribution of areas of hot desert.

1 Describe the distribution of the world's hot deserts.

....

....

....

....

2 Describe the climate associated with hot deserts.

....

....

3 Outline the ways in which vegetation is adapted to conditions in hot deserts.

....

....

....

....

....

4 Outline the ways in which animals are adapted to conditions in hot deserts.

....

....

....

....

....

....

The effects of deforestation

The table below shows some of the uses of tropical rainforests.

Industrial uses	Ecological uses	Subsistence uses
Charcoal	Soil erosion control	Fuelwood
Medicines	Flood control	Fodder for agriculture

1 Add 'Climate regulation', 'Weaving materials and dyes' and 'Tourism' to the above table.

2 Explain why deforestation may lead to increased flooding.

....

....

....

....

3 Explain why deforestation may lead to increased soil erosion.

....

....

....

4 Explain how the composition of soil changes following deforestation.

....

....

....

5 How can deforestation lead to climate change?

....

....

....

....

Climate graphs

1 Using the data on page 156 (Manaus) and page 157 (Cairo) of the textbook, complete the following two climate graphs. The data for January to March have already been plotted.

a

Max temperature

Average temperature

Min temperature

Temperature (°C)

Rainfall (mm)

Jan Feb Mar Apr May Jun Jul Aug Sep Oct Nov Dec

b

Temperature (°C)

Rainfall (mm)

Jan Feb Mar Apr May Jun Jul Aug Sep Oct Nov Dec

2 a State the range of average monthly temperatures in Manaus and Cairo.

Manaus ....

Cairo ....

b Identify the wettest month in Manaus. ....

c Which months have no rainfall in Cairo? ....

3 Compare the average number of sunshine hours in Manaus and Cairo.

....

....

....

4 Suggest why Manaus has a relatively low temperature range whereas Cairo has a high temperature range.

....

....

....

....

Weather maps

1 The diagram below shows the main weather symbols shown on weather maps.

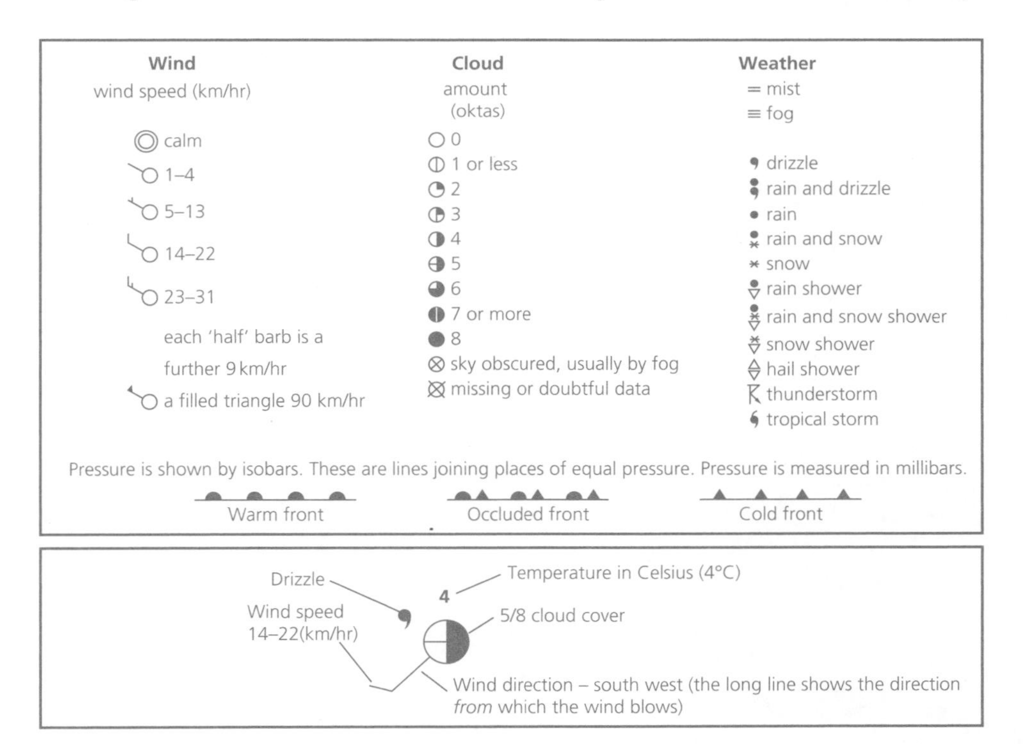

The map below shows the weather conditions for the British Isles during October.

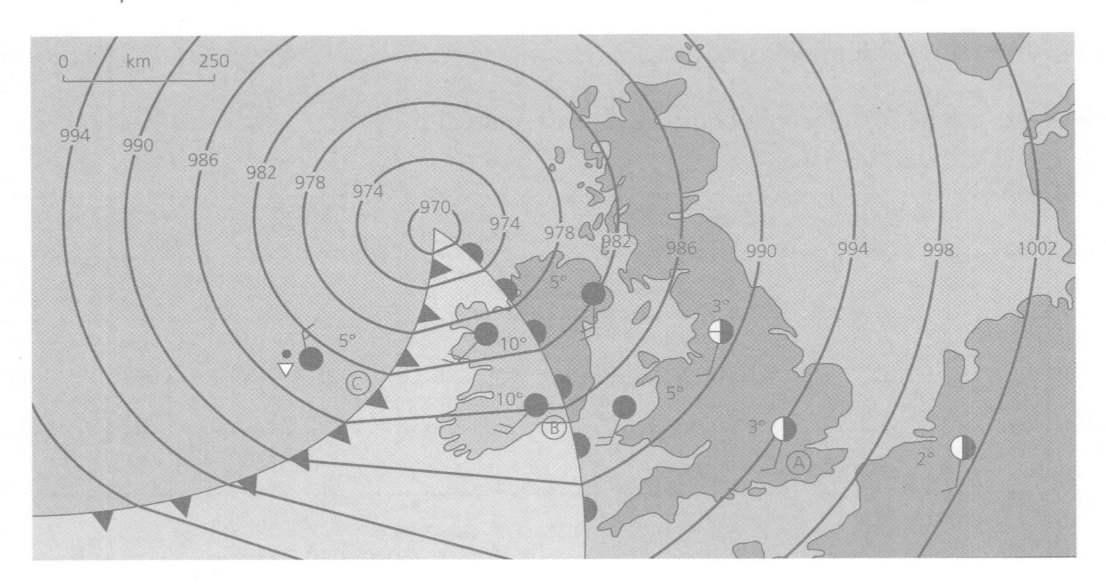

a Identify the type of front located at B. ....

b Identify the type of front nearest to point C. ....

2 Describe the weather conditions at:

A ....

B ....

C ....

3 The following map shows a weather situation during August.

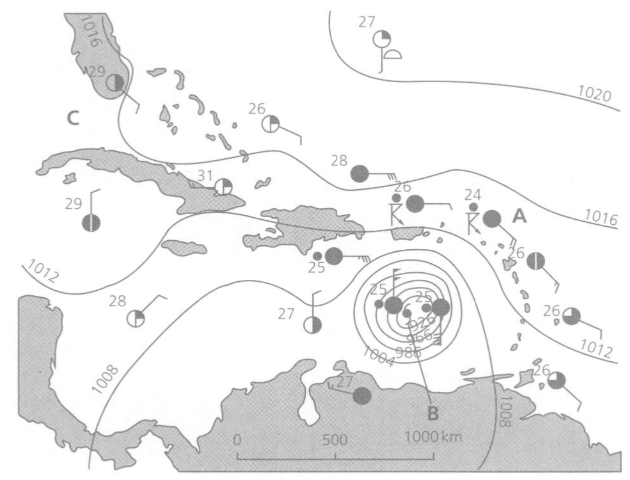

a Identify the weather system at B. ....

b State the air pressure in millibars. ....

4 a Describe the weather conditions at A (St Kitts and Nevis).

....

....

b Describe the weather conditions at B.

....

....

c Describe the weather conditions experienced in Florida, USA — location C.

....

....

3 Economic development

3.1 Development

1 What is meant by the term 'development'?

....

....

....

....

2 Look at the photograph above and also at the photograph on page 172 in the textbook. Both photographs are from southern Mongolia. Describe what the photographs show and how they reflect the level of development in that region.

....

....

....

....

....

....

3 Describe one numerical measure of the level of economic development in a country.

....

....

....

....

4 What is the 'development gap'?

....

....

....

5 Look at Figure 3.3 on page 173 of your textbook. Describe the global variation in GNP per capita in 2013.

....

....

....

....

....

....

6 Why are the following indicators considered to be good measures of development?

a Literacy ....

....

....

b Life expectancy ....

....

....

c Infant mortality ....

....

....

7 List two other individual measures of development.

1 ....

2 ....

8 Why is the human development index a better measure of development than the individual indicators considered in questions 6 and 7?

....

....

....

....

9 List the six countries with the highest levels of human development in 2016.

1 .... **4** ....

2 .... **5** ....

3 .... **6** ....

10 Insert the two labels 'Least developed countries' and 'Newly industrialised countries' into the correct places on the figure below:

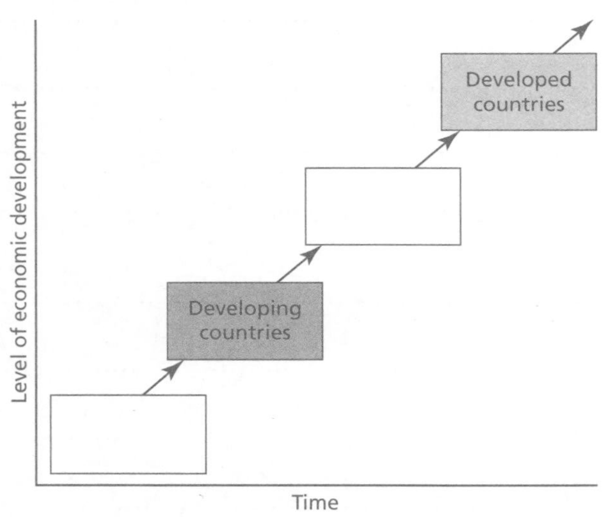

11 Where are the world's least developed countries located?

....

....

....

....

12 Briefly state the reasons for the extremely low incomes of the least developed countries.

....

....

....

....

13 What is a newly industrialised country (NIC)?

....

....

14 Name the first four countries to be recognised as newly industrialised countries.

1 .... **3** ....

2 .... **4** ....

15 Name three other countries that have become NICs more recently.

1 ....

2 ....

3 ....

16 State two aspects of physical geography that have hindered development in various countries.

1 ....

....

2 ....

....

17 Developing countries with good 'institutional quality' have been much more successful than those countries lacking this important development factor. What do you understand by the term 'institutional quality'?

....

....

....

....

18 Name a technique that is frequently used to show the extent of income inequality.

....

19 Look at Figure 3.10 on page 179 in the textbook. Name:

a four countries with very low income inequality

1 ....
3 ....

2 ....
4 ....

b four countries with very high income inequality.

1 ....
3 ....

2 ....
4 ....

20 Insert the two labels 'Economic core' and 'Periphery' in the correct places on the diagram below. Explain the meaning of each of these terms.

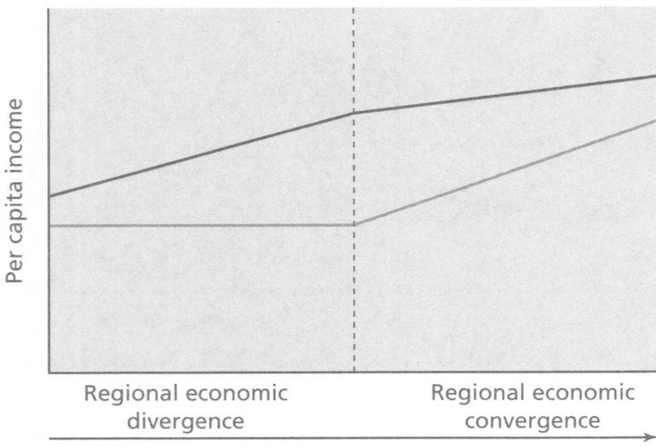

....

....

....

....

21 What is the difference between 'regional economic divergence' and 'regional economic convergence'?

....

....

....

....

22 List four factors affecting inequalities within countries.

1 ....
3 ....

2 ....
4 ....

23 What is the difference between the formal sector and the informal sector within an economy?

24 Look at the photograph below.

Which sector of the economy does the photograph represent? Explain why.

25 Give two examples of jobs in each of the following:

The primary sector

The secondary sector

The tertiary sector

The quaternary sector

26 a Add the labels 'primary', 'secondary' and 'tertiary' in the correct places on the diagram.

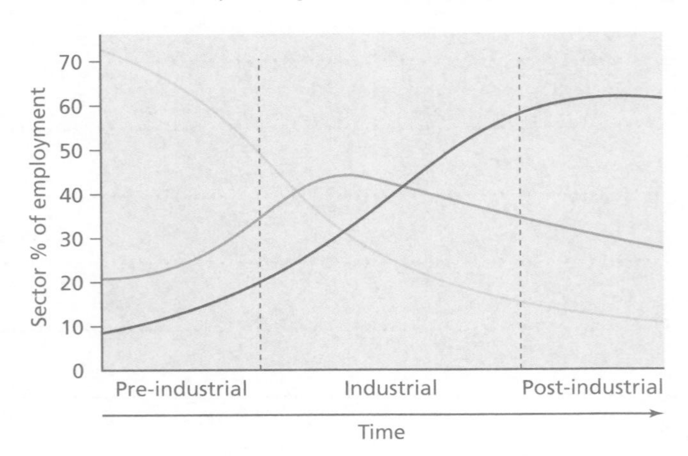

b Explain how these three sectors change in importance over time as an economy develops.

....

....

....

....

....

....

27 Name a type of graph that can be used to show how different countries vary in employment structure.

....

28 Define 'globalisation'.

....

....

29 What is a transnational corporation?

....

....

30 Why have advances in technology been important to the development of globalisation?

....

....

31 State and briefly explain three impacts of globalisation at the global scale.

1

2

3

32 State two impacts of globalisation at the national scale.

1

2

33 Discuss one possible impact of globalisation at the local scale.

3.2 Food production

1 Complete the blank spaces in the paragraph below.

Farming can be seen to operate as a .... with inputs, ...

and .... The cultivation of crops is described as .... farming

while .... farming involves keeping livestock such as ....

and pigs. .... farming involves cultivating crops and keeping livestock together.

2 What is the difference between subsistence farming and commercial farming?

....

....

....

....

3 Complete the right-hand side of the table below to briefly describe the characteristics of the three types of farming.

Type of farming	Farming characteristics
Extensive farming	
Intensive farming	
Organic farming	

4 Give an example of extensive farming.

....

5 Briefly explain three physical factors that influence farming.

1 ....

....

....

2 ....

....

....

3 ....

....

....

6 Complete the blank spaces in the paragraph below.

Economic factors influencing farming include markets, ... and ....
Large farms allow .... of .... to operate, which reduce the unit costs of production. Agricultural .... is the application of advanced techniques in farming. This is particularly important to the improvement of agriculture in developing countries. An important social/cultural factor in farming is land .... The influence of government on farming is classed as a .... factor. An example is the .... in the European Union.

7 What is irrigation?

....

....

8 Which is the most advanced type of irrigation — sub-surface, aerial or surface? ....

9 Define 'agricultural technology'.

....

....

10 What is the term that describes the ways in which land is, or can be, owned?

....

11 List four natural problems that can lead to food shortages.

1 ....

2 ....

3 ....

4 ....

12 Economic and political factors that can also contribute to food shortages include:

1 ....

2 ....

3 ....

13 Why is malnutrition such a significant problem in many poor countries?

....

....

....

....

14 Describe the three types of food aid.

....

....

....

....

....

15 Name the main organisations providing global food aid.

....

....

16 Give one criticism of the way food aid operates.

....

....

....

....

17 The Green Revolution has increased food production significantly in many developing countries.

a State three advantages of the Green Revolution.

1 ....

....

2 ....

....

3 ....

....

b State two disadvantages of the Green Revolution.

1 ....

....

....

....

2 ....

....

....

....

18 Which factor did the United Nations Environment Programme highlight as being essential for significant growth in food supply without compromising environmental sustainability?

....

3.3 Industry

1 In the space provided, draw a simple diagram to show the three components of an industrial system.

2 Explain the difference between processing and assembly industries.

....

....

....

....

3 What are 'footloose industries'?

....

....

4 Is the iron and steel industry an example of heavy industry or light industry? ....

5 What is 'high-technology industry'?

....

....

....

....

6 Name two companies that manufacture high-technology products.

....

7 Where did high-technology industry first develop?

....

8 Why do high-technology industries often cluster together?

....

....

....

....

9 Give one example of a science park. ....

10 Explain two physical factors that influence the location of industry.

1 ....

....

2 ....

....

11 Explain two human factors that influence industrial location.

1 ....

....

2 ....

....

12 What is meant by the term 'industrial agglomeration' and why does it occur?

....

....

....

....

....

....

....

....

13 Define an 'industrial estate'.

....

....

14 What are the reasons for grouping companies together on industrial estates?

....

....

....

....

15 How has the location of industry changed in recent decades:

a on a global scale?

....

....

....

....

b within individual countries?

....

....

....

....

c on an urban scale?

....

....

....

....

3.4 Tourism

1 Define 'tourism'.

....

....

....

2 a Look at Figure 3.56 on page 211 of the textbook and describe the growth in global tourism from 1950 to 2010.

....

....

b What is the projected increase in global tourism between 2010 and 2020?

....

3 What were the reasons for the early development of tourism in the eighteenth and nineteenth centuries?

....

....

....

....

....

....

4 How many passengers did scheduled planes carry in:

a 1970? ....

b 2016? ....

5 Give one example of a political factor influencing tourism.

....

....

6 Define a tourist-generating country.

....

7 In the table below, insert three economic and three social factors that have influenced the growth of global tourism.

Economic factors	1 ....
	2 ....
	3 ....
Social factors	1 ....
	2 ....
	3 ....

8 What is the most important mode of transport for inbound tourism around the world?

....

9 What proportion of the world's exports in goods and services is accounted for by international tourism?

....

10 Which world region accounts for over 50% of international tourist arrivals? ....

11 Why is seasonality a major problem in many tourist destinations?

....

....

....

12 Use examples to distinguish between the direct and indirect economic impact of the tourist industry.

....

....

....

....

....

....

....

13 Give the example of informal sector employment in tourism shown in the textbook.

....

14 Explain the meaning of 'economic leakages'.

....

....

....

....

15 State three economic benefits of tourism.

1 ....

....

....

....

....

2 ....

....

....

....

3 ....

....

....

16 In the table below, insert three positive and three negative social/cultural impacts of tourism.

Positive social and cultural impacts	Negative social and cultural impacts
1	1 ....
....	....
2	2 ....
....	....
3	3 ....
....	....

17 Define 'sustainable tourism'.

....

....

....

18 Suggest two ways in which individual tourists can reduce their 'destination footprint'.

1 ....

....

2 ....

....

19 What is ecotourism?

....

....

....

....

20 In terms of protecting tourist destinations, what is the difference between preservation and conservation?

....

....

....

....

3.5 Energy

1 State the non-renewable sources of energy.

....

....

2 Define 'renewable energy'.

....

....

....

3 What is the 'energy mix' of a country?

....

....

4 Which source of energy is currently the most important globally?

....

5 Why is fuelwood such an important source of energy in the developing world?

....

....

....

....

....

6 What is the 'energy ladder'?

....

....

....

....

7 In the table below identify four advantages and four disadvantages of nuclear power.

Advantages of nuclear power	Disadvantages of nuclear power
1	1 ....
....	....
....	....
2	2 ....
....	....
....	....
3	3 ....
....	....
....	....
4	4 ....
....	....
....	....

8 Why are most countries eager to develop renewable sources of energy?

....

....

....

....

....

....

9 a Why is hydroelectric power considered to be a traditional source of energy?

....

....

b Which four countries account for more than half of the world's HEP generation?

1 .... **3** ....

2 .... **4** ....

c Why is the opportunity for more large-scale HEP development very limited?

....

....

....

d Identify three problems associated with the development of HEP.

1 ....

....

2 ....

....

3 ....

....

10 Look at Figure 3.81 on page 226 of the textbook. Describe the increase in renewable energy consumption shown.

....

....

....

11 a State the worldwide capacity of wind energy.

....

b Which four countries are the global leaders in wind energy?

....

....

c Why have so many countries invested in wind energy?

....

....

d Identify three concerns about the development of wind energy.

1 ....

2 ....

3 ....

12 a What are biofuels?

....

....

b Which two countries are the biggest producers of biofuels?

....

c What are the advantages of biofuels as stated by people who support their production?

....

....

....

....

d What are the disadvantages of biofuel production?

....

....

....

....

13 a Define 'geothermal energy'.

....

....

b Name the leading country using geothermal electricity.

....

c Give three advantages of geothermal energy.

1 ....

....

2 ....

....

3 ....

....

d State three limitations of this form of energy.

1 ....

....

2 ....

....

3 ....

....

14 a What is the global capacity of solar electricity?

....

b Which five countries are the leaders in the production of solar electricity?

....

....

c Describe the two ways in which solar electricity is produced.

1 ....

....

2 ....

....

d Discuss the advantages and disadvantages of solar electricity.

....

....

....

....

....

....

....

....

15 a How can electricity be produced from tides and waves?

....

....

....

....

b Why is so little electricity being currently produced by these methods?

....

....

....

....

3.6 Water

1 Why do water experts refer to a 'global water crisis'?

....

....

....

....

....

2 Define 'water supply'.

....

....

3 Why are dams and reservoirs so important to global water supply?

....

....

....

....

4 On the diagram below, label the following: the aquifer, two areas of impermeable rock, the source of the groundwater, the artesian well that flows naturally.

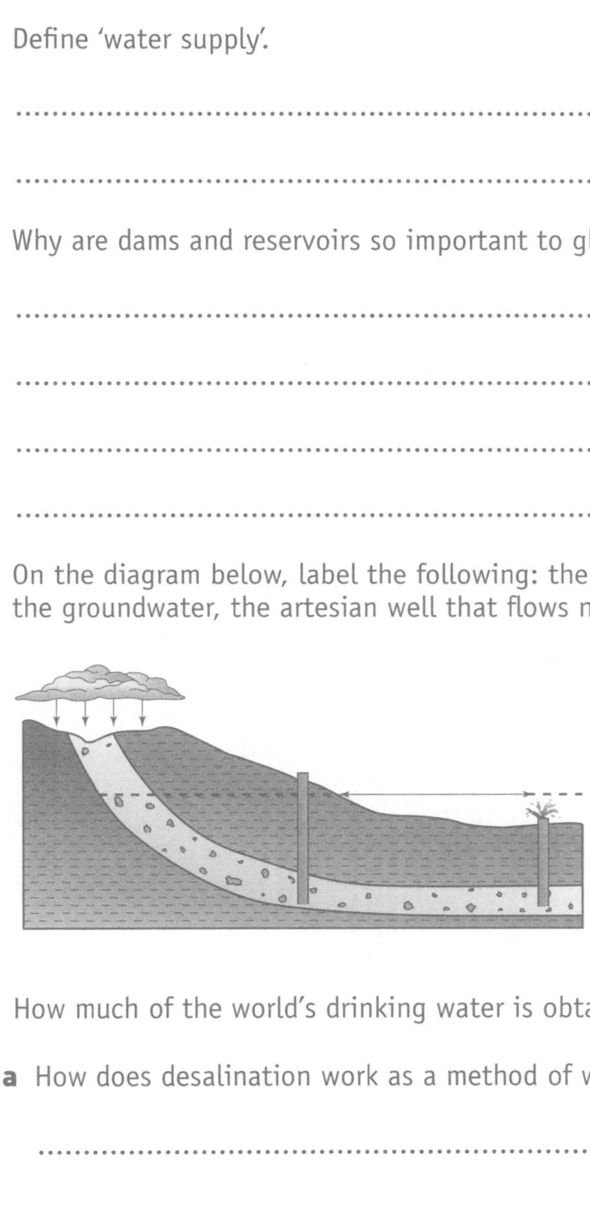

5 How much of the world's drinking water is obtained from aquifers? ....

6 a How does desalination work as a method of water supply?

....

....

....

....

b What are the advantages and disadvantages of desalination?

....

....

....

....

....

7 Name a technique that can be used to increase precipitation in a region. ....

8 Comment on two other methods of water supply that are not covered in the previous questions.

1 ....

....

....

2 ....

....

....

9 Define potable water.

....

10 Look at Figure 3.91 on page 234 of the textbook. Describe and explain the differences in water use shown.

....

....

....

....

....

11 a Explain physical water scarcity.

....

....

....

b Explain economic water scarcity.

....

....

....

12 What is the difference between water stress and water scarcity?

....

....

....

....

13 Why do scientists expect water scarcity to become more severe in the future?

....

....

....

....

....

....

14 How can water management improve the water supply situation?

....

....

....

....

....

....

3.7 Environmental risks of economic development

1 Define 'pollution'.

....

....

....

2 What are the methods of human exposure to pollutants?

....

....

....

3 Which type of pollution has the most widespread effects on human health? ....

4 Name four major air pollutants.

1 ....

2 ....

3 ....

4 ....

5 Define 'externalities'.

....

....

....

6 On the diagram, label the following: the externality gradient, the point of maximum environmental impact, the geographical extent of environmental impact.

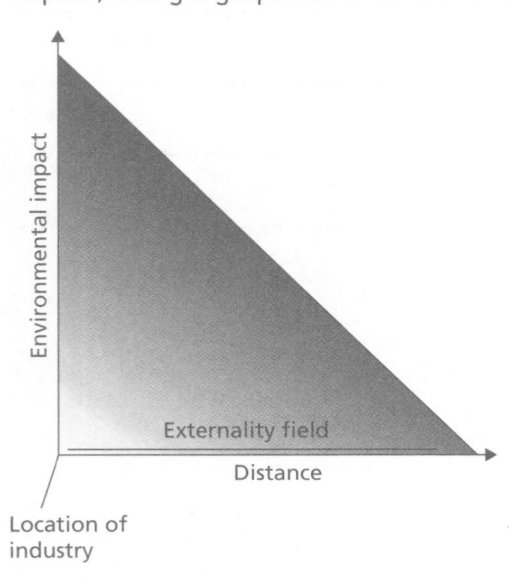

7 How serious is the global water pollution problem?

....

....

....

....

....

....

8 Name a major contributor to noise pollution.

....

9 a Define 'light pollution'.

....

....

b What are the causes and consequences of light pollution?

....

....

....

....

10 Explain the difference between incidental and sustained pollution.

....

....

....

11 Describe one major example of incidental pollution.

....

....

....

....

....

12 Define acid deposition.

13 How is wet deposition formed?

14 State two environmental impacts of acid deposition.

1

2

15 Describe the natural greenhouse effect.

16 How has this effect been 'enhanced' by human activity?

....

....

17 List three of the main greenhouse gases.

1 .... **2** .... **3** ....

18 Discuss three consequences of enhanced global warming.

1 ....

....

....

....

2 ....

....

....

....

....

....

3 ....

....

....

....

19 Define 'soil erosion'.

....

....

20 How serious is the global soil erosion problem?

....

....

....

....

21 Explain the two major causes of soil degradation.

1 ....

....

....

2 ....

....

....

22 Define 'desertification'.

....

....

23 State two physical causes of desertification.

1 ....

2 ....

24 State two human causes of desertification.

1 ....

2 ....

25 Why is soil degradation a threat to food security?

....

....

....

....

....

26 Define:

a 'resource management'

....

....

b 'sustainable development'

....

....

27 Give one example of resource management.

....

28 Define:

a 'recycling'

....

....

b 're-use'

....

....

29 What are the problems associated with landfill?

....

....

....

....

....

....

30 In terms of energy efficiency, define:

a 'carbon credits'

....

....

b 'community energy'

....

....

c 'microgeneration'

....

....

31 Give three examples of ways in individuals can conserve energy.

1 ....

....

2 ....

....

3 ....

....

4 Geographical skills and investigations

4.1 Geographical and mathematical skills

Map skills

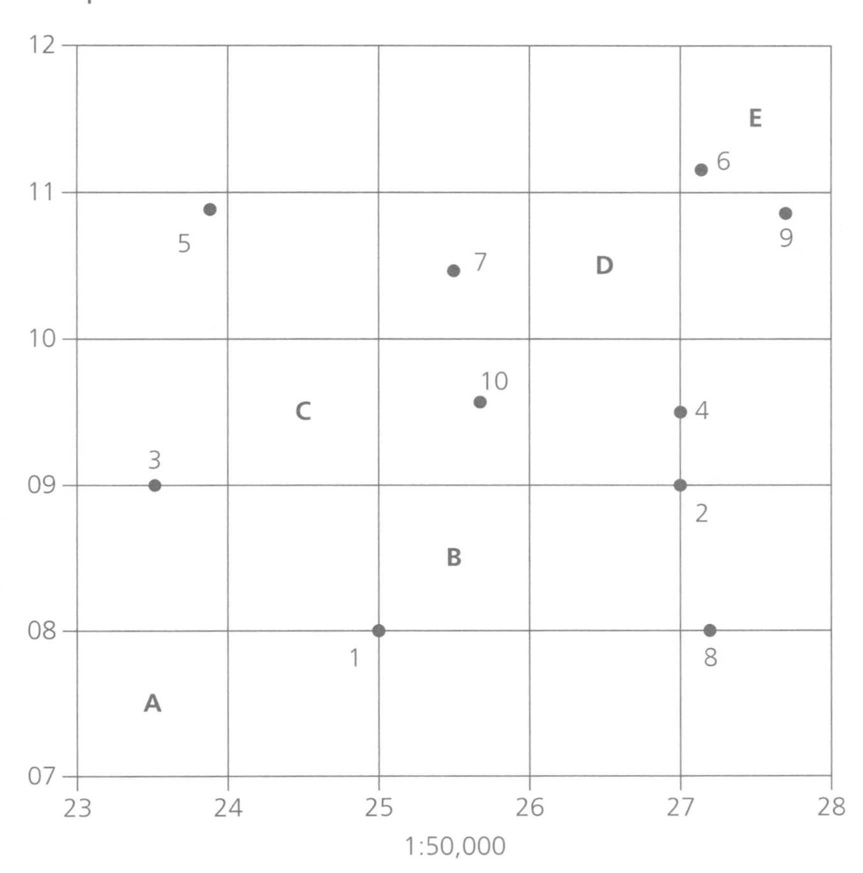

1 State the four-figure square reference for the following squares:

A ....

B ....

C ....

D ....

E ....

2 State the six-figure grid reference for the following points:

1 ....

2 ....

3 ....

4 ....

5 ....

6 ....

7 ....

8 ....

9 ....

10 ....

3 State the distance between the following points:

2 to 4 ....

1 to 8 ....

3 to 2 ....

4 to 9 ....

7 to 2 ....

4 State the direction of:

7 from 10 ....

2 from 4 ....

1 from 8 ....

8 from 1 ....

9 from 1 ....

Mapwork and cross-sections

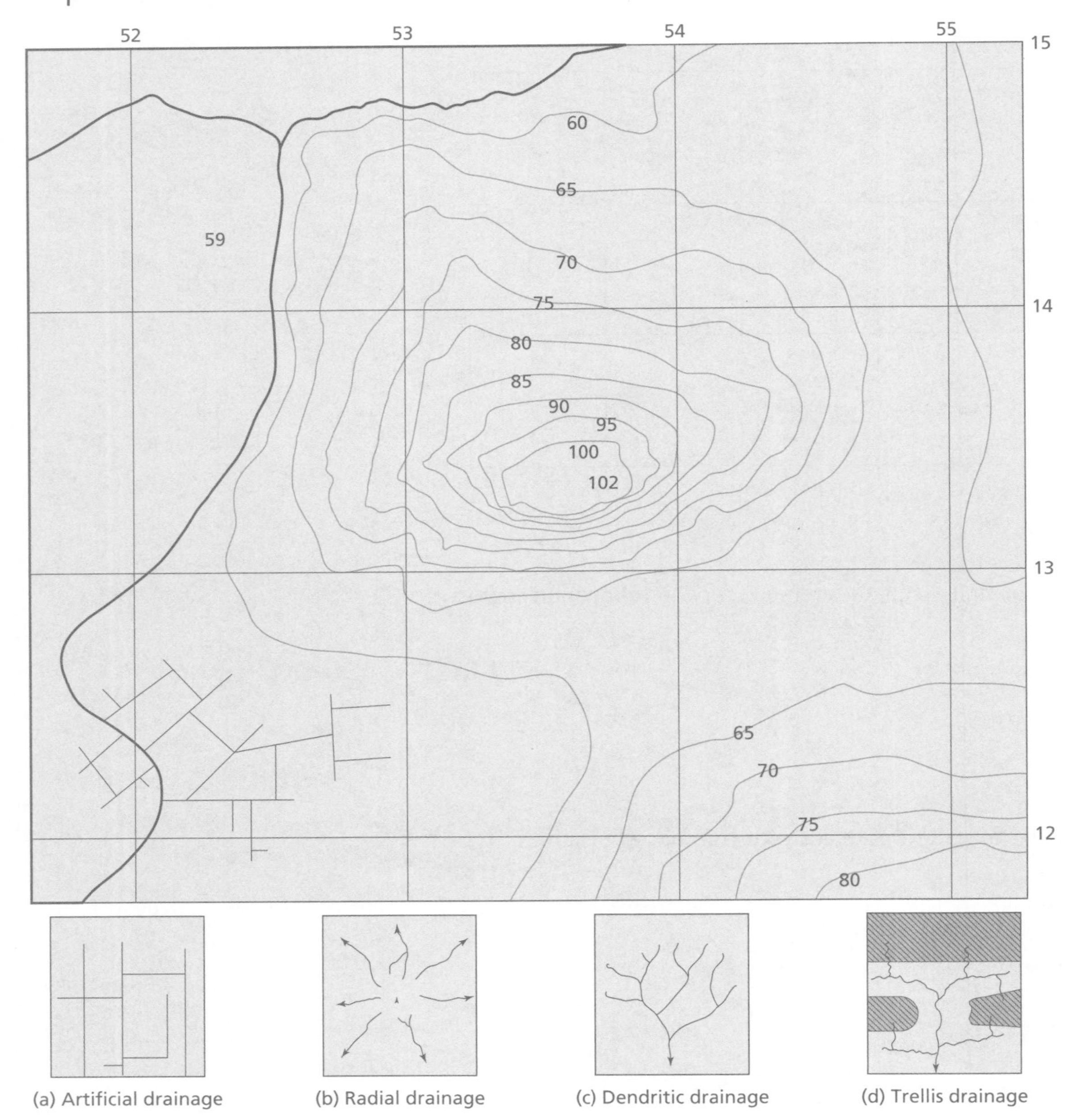

(a) Artificial drainage (b) Radial drainage (c) Dendritic drainage (d) Trellis drainage

1 a Complete the cross-section below from 520150 to 540130 to show the changes in relief and drainage.

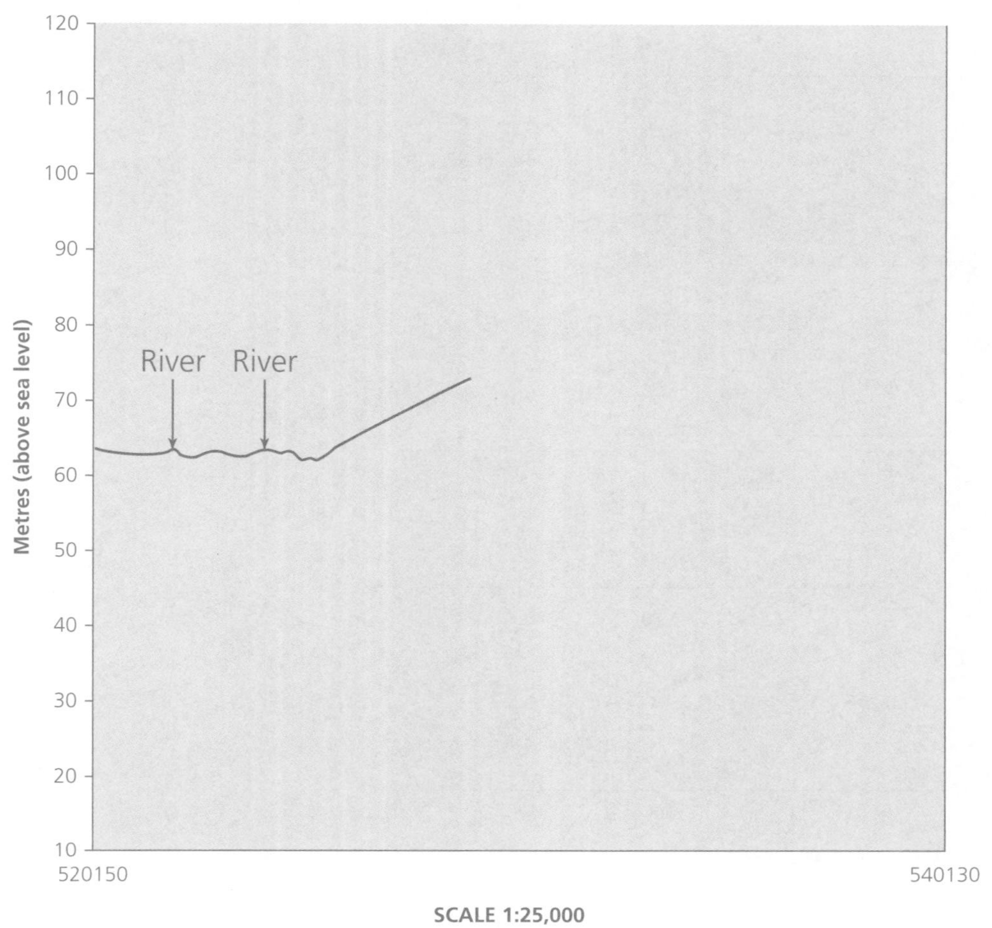

b Identify the type of drainage in square 5212. ....

c Compare the relief of square 5313 with that of 5212.

....

....

d Identify the landform to the west of the river in square 5213. ....

e Identify the feature at 526147. ....

f Compare the relief in square 5313, south of spot height 102 metres, with that of square 5314.

....

....

Valleys and spurs

A river valley is an area of lowland next to a river, while a spur is an area of higher land extending into areas of lower land.

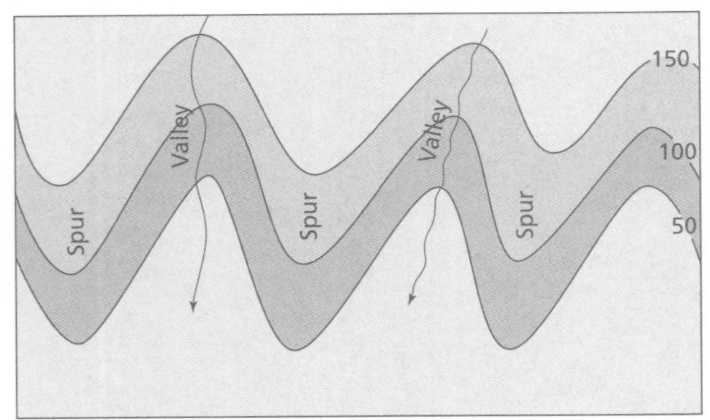

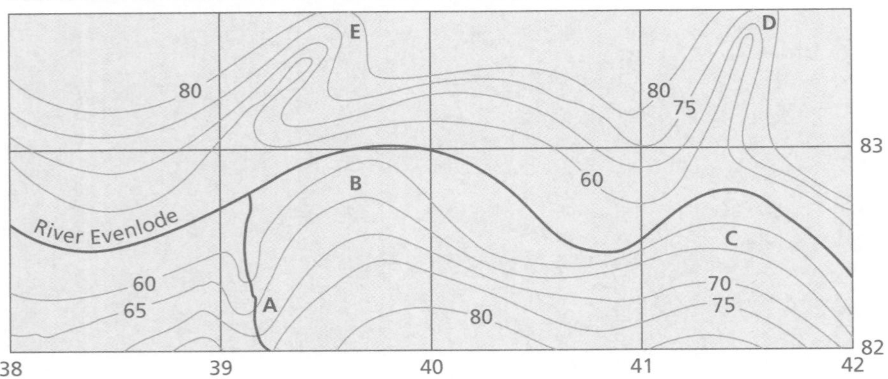

1 Identify the features A to E on the map.

A ....

B ....

C ....

D ....

E ....

2 Suggest a meaning for the term dry valley.

....

3 Identify two dry valleys on the map. ....

4 Describe the valley of the river Evenlode.

....

....

....

....

5 Suggest one advantage and one disadvantage of the Evenlode valley for transport.

....

....

....

....

6 Look at Photo A and suggest why flooding on the Evenlode is extensive.

Photo A

....

....

....

....

7 Identify the feature in Photo B.

Photo B

....

Maths for geographers

Study the map of Woodstock and Blenheim Park on page 282 of the textbook.

1 Estimate the approximate area of Woodstock in squares 4416 and 4417. ....

2 Identify the highest spot height within the built-up area of Woodstock, and state its six-figure grid reference. ....

3 Work out the shortest straight-line distance from the point identified in question 2 to the spot height 87 m near Upper Campsfield Road (463164). ....

4 The surface of the lake is just under 80 m above sea level. Using the rounded figure of 80 m, work out the steepest gradient from the highest point to the lake.

a Work out the gradient between the point identified in question 2 to the spot height 87 m near Upper Campsfield Road (463164). Express your answer as 1:*x*. ....

b State the orientation of the A44. ....

5 The photo shows the Grand Bridge taken from 443167. State the direction that the camera was facing and the time of day when the picture was taken. Justify your answer regarding the time of day.

....

....

Mapwork and photos

Study the map of Woodstock on page 282 of the textbook and the four photos overleaf, all taken from the centre of the Grand Bridge (449164).

A

B

C

D

State the direction that the camera was facing in each of the photos, and explain how you reached your decision.

Photo A

....

....

Photo B

....

....

Photo C

....

....

Photo D

....

....

Analysing photos

Photos A and B were taken from the same location in the northern hemisphere. In Photo A the camera was pointed at 270° and in Photo B the camera was pointed at 225°.

Photo A

Photo B

1 In which compass point directions was the camera pointing for Photos A and B?

A .... B ....

2 At what time (morning, noon, afternoon) were the photos taken? Give evidence to support your answer.

....

....

3 a Describe the distribution of population in Photos A and B.

....

....

....

....

b Suggest a possible reason for the distribution of population that you have described.

....

....

....

....

c Suggest how this may help explain the distribution of population on a global scale.

....

....

....

....

Reinforce learning and deepen understanding of the key concepts covered in the revised syllabus; an ideal course companion or homework book for use throughout the course.

- Develop and strengthen skills and knowledge with a wealth of additional exercises that perfectly supplement the Student's Book.
- Build confidence with extra practice for each lesson to ensure that a topic is thoroughly understood before moving on.
- Improve geographical skills such as data interpretation, and diagram and map reading with practical applications and exercises.
- Keep track of students' work with ready-to-go write-in exercises.
- Save time with all answers available online in the Online Teacher's Guide.

Use with *Cambridge IGCSE™ and O Level Geography 3rd edition*
9781510421363

This resource is endorsed by Cambridge Assessment International Education

- ✓ Provides learner support for the Cambridge IGCSE™ and O Level Geography syllabuses (0460/0976/2217) for examination from 2020
- ✓ Has passed Cambridge International's rigorous quality-assurance process
- ✓ Developed by subject experts
- ✓ For Cambridge schools worldwide

HODDER EDUCATION
www.hoddereducation.com

ISBN 978-1-5104-2138-7